Effective Methods for Documenting the Paranormal

If you follow THEIR RULES

THEIR
RULES
An Interactive Book

June Matthews
&
Laurel Cooper

Copyright © 2022 June Matthews, Laurel Cooper

All rights reserved. The Matthews Family

ISBN: 979-8-9860583-3-7

DEDICATION

Traditionally, this would be a dedication to family, friends, people who inspired us, people who have supported this book, or other people we feel are important enough to mention on the dedication page. This book is for the people who are no longer here in the flesh but in spirit. To those who bang at night, send orbs around your house, speak with disembodied voices, and inspire people to try and capture evidence of the paranormal. This book we dedicate to the dead; after all, these are THEIR RULES

CONTENTS

	IMAGINE	ix	
	INTRODUCTION	Page	01
	HOW IT WORKS	Page	07
RULE #1	BE OPEN-MINDED	Page	11
RULE #2	BETTER PLAY NICE	Page	17
RULE #3	CAREFUL WHAT YOU ASK	Page	21
RULE #4	DON'T SCARE ME	Page	33
RULE #5	USE OUR TOOLS	Page	41
RULE #6	WE WANT TO PLAY WITH THAT	Page	63
RULE #7	ARE WE WHAT YOU EXPECTED?	Page	73
RULE #8	WE CAN BE ANYWHERE	Page	79
	ABOUT THE AUTHORS	Page	91
	STEP BY STEP	Page	99
	NOTES	Page	103
	THE PARANORMAL IS REAL	Page	105

IMAGINE

Imagine you find yourself in a strange house and do not know how you got here. You are scared.

You know something happened, but you can not remember what. The most frightening of all, you don't even remember your name. Suddenly you see someone. The woman you see is wearing a long nightgown. She is smiling, not at you, but her smile puts you at ease. You stare at her for a while, hoping she will look your way. She does not seem to know you are in the room with her. Can she help me? You wonder.

You make a slight sound attempting to get her attention. Nothing, she does not even hear it. It takes all your energy, but you try again -still nothing. You do not understand why, but you have to wait before trying again.

Oh, no, she is walking away. You grab at her nightgown; you think this will get her attention. You are shocked when she screams. Why did she get scared? I am not trying to hurt her.

 "Please. Please help me!"

You are unclear how much time has gone by, but suddenly things look different. The lady is back, and another person is standing with her. They start taking pictures. Suddenly they scream at you. "Leave this house, demon!" They talk amongst themselves and refer to you as an evil spirit. They are demanding you leave at once. You are so confused and scared.

"I'm sorry, I'm so sorry," you cry.
You drop to the floor sobbing.
"I just wanted help to find my Mom. I'm scared, and I am only seven years old. Help me, please; where is my Mom? "

Photo from 2015
Automatic Night Vision Security Camera

What you just read is a compilation of stories. We never know what is actually happening when we are dealing with the unknown.

I have my guests read "IMAGINE" before we sit down to record, it helps to create the right state of mind you need to communicate successfully.

Close up 2015 photo
My back was hurting but I never felt the pull on my robe
and I did not know this even happened until I reviewed the pictures.

INTRODUCTION

Eleven years ago, my husband and I moved into a large custom home deep in the north woods of Washington state. It took me almost three years to accept that our home was what some refer to as a Haunted House. When unexplained phenomena started making me question my sanity, I began to keep a daily record of occurrences, experiences, or observations. That journal was a cathartic process and eventually became THE RELIVING book. In 2016, I woke in the middle of the night feeling strange. I went to the restroom, intending to splash cold water on my face; before I crossed the doorway, I felt myself begin to fall to the ground. I woke up on the floor with my husband Jim over me. He took me to the hospital, and we found out that I had a heart attack. Once released from the hospital, I fell in the same manner from a second heart attack.

My book THE RELIVING, A True Story, was still with the editor, waiting for edit approvals before publishing. With all the medications and my weak condition, I could not read the edits, and I approved the book as it was. The book modifications still had some problems, but the E-Book version was worse.

The recovery process was long, and it was impossible to write. I was able to continue documenting the activity in our home from bed. During this time, I understood that I could only make contact (with The Unseen people) when I followed particular guidelines. I could obtain answers to questions as long as I had not talked to this person before. I needed to supply a variety of ways to answer the questions. The Rem-Pod worked one day but not the next.

I learned early on that recording electronic voice phenomena known as EVPs took at least twelve minutes before any EVP voice was recorded. I knew early on that those in my house were not evil or scary. I wrote THEIR RULES because I wanted people to know that the evil representation I see in movies or television programs was false. I felt it necessary also to bring those who claim to be experts in this field to the surface. There can be no experts in a world of the unknown.
I would watch the experts blatantly announce, "This is evil." Of course, anyone may have an opinion, but some of the investigators on TV make factual statements. "I feel the devil here!"
 I wrote about my research to let those who still believed in the dramatic media view that these misguided explanations of the paranormal are for ratings; not everything is evil, and not all shows are like that. One of my favorite shows would carefully tell homeowners who claimed they felt demons in their house that activity is rarely evil, only misunderstood.

By 2020 I had completed this book. The virus and stay-home orders delayed the publishing for a year.
 In October 2020, our daughter Laurel needed to come and finish the lockdown at our home. It was a very emotional time for all of us. To keep ourselves busy, we documented the house almost daily. Within months, I learned that I had many misconceptions about this world.

I had no idea what I was thinking, but before this time, the idea that they could hear me after the recorder was off didn't occur to me. I was wrong.

In the first version of this book, I did not realize that being open-minded mattered as much. Open-Mindedness also included removing sections of the previous version (of this book) that contained my opinions and ego.

The questions we ask should not be the questions we see on TV. Asking someone, "When did you die?" is a question that carries responsibility,
I would ask this question only to have all the equipment suddenly stop. You could actually feel the room empty out.
I asked, "What is your name?" every time and had very few responses from EVPs or devices like the Ovilus. Laurel and I started to understand how difficult these questions can be for some of them.
This book needed to be re-written to include all the critical information we learned.

Since I moved into this home, there is not one thing I can say resembles any movie I have ever seen. Many of the events in this house I had never heard of before. For instance, our light bulbs unscrew themselves; batteries will go dead and then come back to full strength.

In movies or television shows, the number of misconceptions is insane. In films, the Halloween season usually conjures up ghosts; the opposite is true in our home. We have never been successful in recording EVP with scary Halloween décor in the house.

We have seen many unexplainable and strange things here, but we have learned even more. We want to help others understand the reality of what to expect.
In the last ten years, I have learned that those who reside in our home with us are frightened, confused, and crying out for help.

Empathy towards the dead was the first RULE we learned here. In this book re-write, I also explain that although my house is not dangerous or evil, maybe another place or location is. Many people believe that anything paranormal is from satan. The unseen people in our home are not harmful and not the devil. However, this is my experience, and I have no idea if every house is benign.

The most important advice I can stress is to trust your inner voice. If it feels unsafe to you, STOP. No one can know who you are speaking with; I believe God is all-powerful, and if there are demons, they will not affect us. I know that intention matters. If you want to talk to evil, perhaps you will. If you desire to help them, you will.
After almost eleven years now, living here has been a positive experience; the unseen people in our home are only scared of the unknown.

OUR MEMBERS:
The group is for people interested in the paranormal but don't know where to start, may have activity in their homes, and feel they cannot say anything for fear of judgment.

We want to show others the reality of the paranormal and inform them about an entire community of believers that once shared your current worries.

Why do you need to become a member?
This is an interactive book; QR codes are located inside to open videos of communication sessions and more.
Members have access to current and FUTURE recordings and information. The QR Codes open pages that are hidden from the public, and only the book owners may have access.

All Members sign in and create a profile. You decide what to add to your profile. Most members use nicknames, and all information is private. You are not required to add anything but an actual email address and a first name; even using a number is okay. Members have no access to any information regarding other members. Members can interact with each other and share information.

Once a member, you will have access to:

Inexpensive alternatives for high-priced equipment.

Videos of the equipment in action.

Sample photos, video, and audio files, of documented paranormal activity.

Help with your home investigation from other members and us

Live Events and Contests

Learn what to expect and why you must follow what we have learned to call THEIR RULES.

HOW IT WORKS

WELCOME, and thank you for your purchase of THEIR RULES.

Okay, you have bought the book; let us explain how this works:

STEP 1: Log on to the website: www.TheirRules.com.

If you are already a member, skip to STEP THREE:
Non-members will view the Website in GUEST-VIEW MODE. "GUEST-VIEW" MODE Does not Require a QR Code to open

STEP 2: Go to the Community link directly under the book and candle. Read about joining the group. It's so safe & easy.
You will be asked to create a user name and password: Add your name or any name you like. You will control what you add and what you don't.

STEP 3: As you read the book, keep your phone or Ipad handy. You will see sections of the book with QR Codes.

Scaled-down Example:

WATCH: Ghost Box Part Two
NEW QR CODE

Grab your phone and scan the code.
The QR Codes will open hidden pages with videos, photographs,

and audio files pertinent to what you are reading.

STEP 4: We have so much recorded evidence; some pages contain up to 6 videos.

WATCH: HIDING UNDER TABLE
NEW QR CODE- KEEP OPEN

When you see - Keep Open, you do not have to keep it open. This is only for your convenience, so you do not have to scan as many times.

STEP 5: Once you have opened the page with the QR Code
At the top of the webpage, you will find a LINK MENU, look for the name of the link from the book and click or tap it to go to the specified video or photograph.

When you see NEW QR Code =This means another scan is needed to open a new page
When you see NEW QR Code- Keep Open= additional videos will be on the webpage.
When you see SAME QR Code =Same as before but still scanable.

HAVE FUN ON YOUR INTERACTIVE PARANORMAL JOURNEY!

Follow THEIR RULES

RULE #1
BE OPEN-MINDED

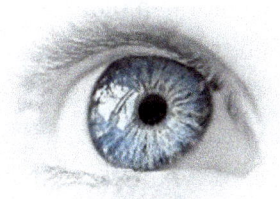

Open-mindedness is a characteristic that involves being receptive to a wide variety of ideas, arguments, and information. Being open-minded is generally considered a positive quality, and it is a necessary ability for critical thinking.
There are a few different aspects to open-mindedness: In everyday use, the term open-minded may define someone as non-prejudiced or tolerant. From a psychological perspective, open-mindedness can describe how willing a person is to try out new experiences. Open-mindedness can also involve asking questions and searching for information that challenges your beliefs.

Since moving here, I realized I had experienced many paranormal events in my life. At the time of those events, I just dismissed them as coincidence. I was closed-minded that anything else was possible, yet I would have told you, "Oh yes, I am very open-minded."

Being Open-Minded is not as easy as you might believe. If you ask almost anyone you know if they are open-minded, they will answer YES; I know many closed-minded people who claim, "I never judge anyone or anything.!" but yes, they do.

I would never recommend doing a recording alone. Of course, you can turn on a recorder and try to capture proof of the paranormal by yourself. It is best not to attempt this alone; why not?
The answer may not be what you suspect. The one thing there is to fear is yourself. If the sound of knocking behind you startles you so significantly you jump from your chair and break a leg, you will be grateful you are not alone.

Attempting this alone can induce the fear response that many people have seen on television and in the movies. We have all seen shows and movies where horrifying things jump out of the darkness and attack you. When I first started this, I was not alone, and if I had been, I could see myself becoming frightened enough that I may have run out of the room. The minute you turn out the lights, all the paranormal movies you have seen will pop up in your mind.

What does this have to do with being minded?
Understanding and accepting our limitations is being Open-Minded. Try to stay open to the idea that you may not know how you will react when you hear your first disembodied voice. So, before you start your recordings, invite over friends. One friend is good; a group would be better. The more people participate in your

attempt to communicate, the more your chances of solid EVPs. I have often had a guest who seems to attract more energy than myself and the others.

Always invite friends you laugh with and enjoy their company. To create the right conditions for recording EVPs and documenting the paranormal, I am a firm believer in the universal truth:

"What you Focus on Expands."

If your guest has a negative complaining personality, you most likely will not succeed in recording EVPs. Remember that you are not just inviting your living friends but attempting to make new friends you have may have never met. If you have a judgmental friend regarding the spiritual, we suggest leaving that person out.

Be careful who enters your home. I had a guest who brought a friend; I thought that was fine. While recording questions, she exhibited such disturbing behavior that I snuck away to retrieve my handgun. (just in case) We later found out this person had a schizophrenic type of mental disorder. Guests in our house that are mean become meaner, with mental issues become sicker, with anger become angrier. However, the nice become nicest, the loving love more, and these empathetic personalities attract the paranormal.

There is no way to know how people will react or what will happen when dealing with the paranormal your house.

I have had guests with some of the strangest responses to the supernatural. Once a guest just got up and left, destroying the recording for the entire night. Another guest was so scared she could not stop nervously talking and squirming in her seat, also eliminating that recording.

People are people and will always react in unexpected ways.
Be Patient.

The successful EVPs are from when people have been open and honest. One guest began talking about child sexual abuse she had suffered; then, the house became more active, and the recordings were more evident than ever. The responses (EVP Recordings) were from children who related to her experiences of abuse. Due to the nature of those recordings, we deleted them.

We recommend the following:
Invite a friend or friends who are open-minded
Friends you trust
Friends who are not aggressive
Friends who you believe may have sensitivities to the paranormal
People who have strange experiences they don't understand

Ask everyone to bring one meaningful thing to be placed on a plate near the non-stop recorder.

It is difficult to change the way we think about something. Our beliefs are very personal, and many people can change their views. In my time as a drug and alcohol counselor, I watched many clients change everything they believed for the sake of their own lives. Previously, I had been taught that the devil had immense power, and I grew up with people around me believing everything paranormal was demonic.

After changing your feelings about the paranormal, remember you are still a human; you will feel scared sometimes. Take some deep breaths and try to evaluate what is happening. I practiced this daily for years. When logically I can say I am safe, I continued.
However, if you feel unsafe or so frightened at any time, you can not calm yourself, stop.
Announce to everyone (including the Unseen People) that you are uncomfortable with this, and you will stop.

I have never had an experience that required me to stop a session out of fear. I did stop a recording once because I was angry, and I believe I was recording a prankster personality of a teenager. I was feeling sick and made as many mistakes with questions. I wanted to quit and should have. He had started asking for help, and I kept the communication open. Although I do not think he was malicious, I did realize he was lying. I became angry because I felt so sick, and he was playing games.

The reality is there is nothing wrong with setting boundaries and saying, "If you need help, we will have to talk later. I am feeling sick, and cant help you now."

The most important thing to consider is that no one knows what is happening in the afterlife—so understanding that hopefully helps you remember to be respectful, kind, and empathetic. You will not know who you may be talking with, so always be nice.

BETTER PLAY NICE

Before starting any recording session, make sure you are in the right state of mind.

WARNING LIST
IF YOU or GUEST:

Has received terrible news,
Recently argued with someone (on the same day)
Are feeling sick
Are feeling depressed
Not had enough sleep
Have been drinking
Are using drugs

Postpone the recording until you are feeling grounded mentally and physically.

I would suggest not drinking before the communication. If you want to serve alcohol at your gathering, we recommend waiting until AFTER you complete the session.

It helps to take some deep cleansing breaths before you start. Introduce yourself speaking in a pleasant voice and have those in the room with you do the same.

Remember, you may be talking to children. Try not to swear or use any aggressive language.

You want to keep your voice calm and non-threatening. Demonstrating CHILD-LIKE (not childish) behavior gets the best results. Think of the personality traits of the people who make you feel safe and loved, then imitate them. You may be talking to an adult, but there can always be a child close enough to hear you. That's what it is like in our house. We cannot know how many children there are, but we know they are always close from past recordings.

Capturing actual documentation of the paranormal is nothing like what you see in the media. Many ghost hunting shows have an angry host yelling at the environment, "Show Yourself!"
Can you imagine meeting someone for the first time, and they scream at you,
"WHY ARE YOU HERE! WHAT DO YOU WANT?"

Talk to your guest/guests in advance.
Tell them not to use sarcasm such as
"Hey ghost, where are you hiding?"
"Come out and play, Mr. Spooky."

Acting or speaking sincerely and honestly, rather than joking or halfhearted, is needed, although having fun is great.

The happier, the better. Many people say that fear triggers the paranormal. In my research, I have seen that it is *all* heightened emotion. Happiness and laughing will attract the kind of paranormal faster than being too serious. In our home, being solemn scares them, we have more children than adults, and I wonder if being serious reminds them of a possible damaging childhood they had in life.

So have fun. Watch us having fun and see how they react.

WATCH: HAVING FUN
SCAN THIS NEW QR CODE:

Speak with respect the entire time. You want to make contact, not insult or make anyone angry. You have no idea who you are talking with,
Rule # 2 is the most important rule!

RULE #3
CAREFUL WHAT YOU ASK

What are the right questions to ask someone who is in spirit? Think about this you are trying to communicate with someone you have never met. When you meet a new person, how do you speak to them?
Do you start like this?

"Hello, are you friendly or a demon?
How did you get here?
Do you know you're dead?
Are you alone here?
Why are you here?
When did you die?
What do you want here?

So, if this is how you make friends, you don't have any.

The goal is first to become friends and make sure they feel safe. Later you can ask questions. Keep in mind that there is a vast difference between investigating an abandoned building and your (or someone else's) home.

You never want to create any negative vibes or atmosphere where you live. Again you have no idea who you are talking with, so always show respect.

SPEAK OUT:

When speaking out, use a calm and salubrious voice, but loud. It might take practice for some, it did for me. When many of us Speak loud; it is when we are angry or feeling negative.

When I record an EVP, it's usually quiet. One of my friends, Sandy, remarked that the voices we had captured together sounded like a wall of water stood between us. If we can not hear them, they most likely have the same trouble hearing us.

Speak Loud – but in a pleasant tone, never yell.

WHAT YOU WILL NEED:

The Website; TheirRules.com
Membership on the Website
At least one other person - more is better
2 Digital Recorders - Use digital-only, not cell phones
Cameras - Cell phones will work great
Headphones- noise canceling works best
Printed Questions (available free in this book)
Flashlight or Candles
30 Second Timer

TIMING:

Before you ask anything, make sure your timer is next to you.

Waiting 30 seconds after each question is crucial for recording a response. You will want to kick yourself if you interrupt an EVP. Also, make sure you have the two recorders if possible. One device can record non-stop, and the second (if possible) for auto-playback after questions. Do not use cell phones for EVP recordings. Cell phones can be used as cameras and video recorders if muted. Links to free and easy timers is in Chapter: STEP BY STEP

EXPLAIN THE EQUIPMENT:
If you have the equipment, you will need to explain what each device looks like, where it is, and what it does.
For instance, if you have purchased or own a Rem-Pod

"On the _____ is a round device that has a silver stick coming out of the top; you can use it to answer YES to any question. It will light up with lots of colors and makes beeping noises. The Rem-Pod will not hurt you in any way; it is only loud."

Please make sure you explain where all equipment is and what it does. If you frighten someone, that someone may leave.
You are attempting to create trust, and you can start this by warning of things like loud sounds and lights set off by the paranormal equipment.
Explained in detail in Rule #5 (TheirRules website GuestMode)
Keep your explanations simple, using words that children will understand.

DON'T CALL OUT: Never ask to contact someone you know.
Do not call out passed loved ones or specific people; here's why.
If you are in spirit, years have gone by, and suddenly someone starts trying to communicate, "Aunt Sally, are you here?"
Out of desperation, I feel anyone in that situation would respond;
"yes, it's me,"
or
"I'm here."
The intent may not be malicious, just desperate.

Do not ask, "are you a demon?"
Have you ever talked to a teenager who thinks scaring people is fun? Remember, these are people who will behave in the afterlife as they did in life.

Inform your guests not to speak with passed loved ones.
If there is no response at all, what does it do to you or your guest emotionally when the specific "loved one" does not show up?
Worse would be the teen prankster responding and sending your guest into an emotional frenzy.
If you have a friend who is experiencing loss it is wise to NOT include them in this event at all.
The psychological effects of grief can be serious.

If nothing happens at your event (party), be patient and do not instigate negativity. "Hey, are you a coward?"

If something happens, and someone overreacts, is afraid, and starts screaming, stay calm and remind them nothing has happened. We can all be frightened of the unknown, and the media has conditioned us to respond in this manner.
Speak logically; I always use "There was a loud noise, knives didn't fly through the air."

Guest talking about abuse may result in amazing EVPs, but that person is also entitled to their privacy. If you plan on sharing the recordings, make sure your guest is okay with sharing.

TIMING MATTERS: After you have finished explaining Equipment and Trigger Objects, you will be near the 12 Minute Mark.

It may be different in your home; we have never recorded an EVP before twelve minutes. I'm not saying it must be precisely twelve minutes, but it seems some time must pass before getting results. I feel they are simply checking us out, watching and listening, trying to determine what we are all about—making this next part important in your success. Of course, EVP voices often respond to direct questions, but that's after they become comfortable doing so. Now introduce each guest one at a time and start a conversation about each person.

EXAMPLE:
"This is my friend Lucy; we met in High School and have been best friends ever since. She likes Pizza and eats it all the time."

Encourage your friend/friends to chime in. Have a conversation. If someone is with you during your talk, they, more often than not, will comment on what you're sharing. It may feel silly doing this, and that's okay. It will feel staged, but it can be a quick way for your unseen guest to feel a part of the group.

"Oh, Carolyn, you like Pizza, right?'

"Oh yeah, I love it Hawaiian style with pineapple."

"So do I."

"Does everyone here like Hawaiian Pizza?"
You may hear a faint "yuck" on playback.
Introduce each guest in this same way. Give each participant at least three to five minutes to discuss likes and dislikes. We often sat around just talking and later discovered we recorded an unseen person agreeing with our topic.

You should now be at the 20-30 Minute mark. It is a perfect time to start:
(QR CODES FOR ONLINE and PRINTABLE QUESTIONS at the end of this chapter)

THE HOST QUESTION # 1
"Again, my name is _____. We want to know about you. Will you please talk with us?"

Wait 30 Seconds
THE HOST QUESTION # 2
"Will you please tell us what name you would like us to call you?"

NO NAMES: If you have written out your own Questions List, ask in the same manner as above.
Don't format questions like: "What is your name?" - "Tell me your name."
When asking their name try to keep this in mind. I'm not sure why (I have my theories), but many of the people in our home don't always seem to know their names. I always try to ask, "What may we call you, or what name would you like us to call you?"
My theory is that this could be from the trauma they experienced right before death.

Many years ago, I had a bad reaction to medication and passed out in my building's underground garage. Other tenants from the apartment complex called 911. When the paramedic revived me, I was shocked that I had passed out and hit my head on concrete. My head hurt terribly, but otherwise, I was fine. Then the paramedic asked me what my name was; I became hysterical. I had no idea that I did not know who I was until he asked.

Over the ten to eleven years we have been documenting this house, I spent the first eight years asking, "What is your name?" or "Please tell us your name," and I have very few responses from EVPs. Many times, the entire recording session came to a halt, and equipment and recorded voices stopped responding entirely.

The same results with questions such as:

"Did you die in this house?"
"When did you die?"
"How did you die?"

That paramedic did not ask me any more questions about my name, only can you move your fingers, feet, etc. If he had asked me any questions solidifying, I had amnesia; I would have become terrified. Thank God he knew what he was doing. If he questioned, "Are you female?" I would have looked at my body before I answered.

If you have ever had amnesia, you know what I mean. It was as if I knew nothing at all about myself, where I was, who I was, or was I male or female.

What if he had asked, "Is the fall what killed you?"

Can you imagine how that would feel?

So, there must be a possibility that some don't know they are dead at all, and I don't want to be the one to tell them. I change that type of question to this: "Do you remember what happened to you?"

That typically gets a simple yes or no response; after that, I don't push it. I have also stopped asking gender questions directly.

MONITORING EVP RESPONSES: Assign one guest to listen to the playback recorder, to monitor responses. The reason for this may be different than you think—the list of questions: is designed to elicit a response. Once you have an answer to your question, your following questions should correlate to that answer. Resembling more of a conversation; otherwise, it is confusing and can sound horrible. Here's what we mean.

Not Monitoring Responses:
Question: "Do you feel safe here?"
EVP Response: "No"
Question: "So you are happy here?"
At this point, if you were having this conversation with someone, you would most likely leave the room.

Monitoring Responses:
Question: "Do you feel safe?"
EVP Response: "No"
Question: "I'm sorry that you don't feel safe. Do we make you feel unsafe, or is it something else?"
EVP Response: "You"
Question: "You are safe with us, but we understand you just met us; how can we help you feel safe?"

Unclear Responses:
You hear a response, but it is unclear what was said. <u>Do not guess.!</u> If some in your group hear Yes and others hear No, this could be the outcome:
Question Asked: "Are you happy here?"
EVP Response: "Yes"
List Question: "Why are you not happy?"

It is better to reply with an honest answer such as: "We did hear you respond, but we still can't make out what you said. We want you to be happy here, and we are happy to be talking with you."

QUESTIONS: All the questions are in the present tense. We recommend reading this book in its entirety before starting. If you ask a question and have no response, have someone else ask the same question.

OPEN ONLINE QUESTIONS:
SCAN THIS NEW QR CODE:

Helping to ensure your guests can easily see the questions. An iPhone may work but may be too small to read. Use an iPad or computer during your session for a lighting alternative. Open the above QR Link; Host Copy is the first scan down the page for each guest.

PRINT OUT QUESTIONS:
This webpage contains printable questions.
SCAN THIS NEW QR CODE:

NOTES: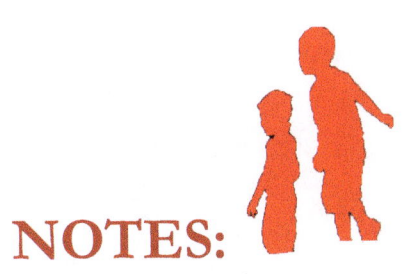

RULE #4
DON'T SCARE ME

Hopefully, your journey into the paranormal is to understand, communicate and help if needed. Just as you do when you invite guests to your home, create a pleasant atmosphere.

As I mentioned in the Introduction, we have never successfully recorded EVPs when our home has scary Halloween decorations. We have a yearly Halloween Party; my guests want to sit for (what we call) a séance. It is not a séance; by definition, it is a group of friends sitting together, asking questions while monitoring and recording the responses.

In the beginning, I still believed that fear would increase our chances of a successful night (I learned that nonsense on TV.) Our Halloween decorations became more frightening as each year went by. It took a while to realize that the atmosphere was too creepy; guess who we scared away?

We record more children here than adults. Some of the scary animatronics we have are fun for Halloween parties; children would likely run screaming. Since you can not know the ages of the entities inside your home, we suggest friendly décor in the area you will be documenting. We now put the scary stuff outside.

We have also discovered some essential things to consider. Language like swearing can drive away a child's spirit in our home.

We have neighbors who target practice on their property and

depending on their weapons, we can hear the gunfire inside the house. One afternoon Laurel and I were documenting downstairs. We listened to the shots, the SLS Camera figures disappeared. I accidentally dropped an EMF detector that went off. I held the SLS camera under the table and found one child spirit hiding under the table, terrified to come out.

WATCH: HIDING UNDER THE TABLE
SCAN THIS NEW QR CODE:

Screaming, yelling, dogs barking, and loud unexpected sounds can quickly stop EVPs and devices. When recording EVPs, we suggest that you wait until night. Of course, you can record any time of day; the paranormal does not need the dark. However, outside interference is louder during the day, depending on your location. Researchers and investigators often record EVPs at night since it is less likely to register daytime noises.

If you are using the type of recorder we will recommend in the equipment section of this book; you are recording sounds almost ten times louder than the human ear. So everyday sounds can re-play as if the devil is in your house.

If you are looking for items such as the tablecloth below, there are many items on the website TheirRules.com. Two Collections

THEIR RULES Collection (tablecloth) and CROW & MOON Collection.

PREPARE THE SPACE:
Serve food an hour or more before recording. Gurgling stomachs and restroom breaks can disrupt any results.
Serve non-gassy foods if possible; this is self-explanatory

Turn off things that create sounds, clocks, air conditioning, fans, house phone and have guests shut off ringers on their cell phones.

A round table works best - easier to pass equipment you can see everyone.

The space you will be recording in is not relevant; it's the atmosphere you create. Just as you would set up for guests in your home, set up for the guests you will be attempting to meet.

Items with words like "Ghost Hunting" should be removed.

Set up the room so that ALL your guests are comfortable. While recording a guest moving in their chair can sound incredibly confusing on playback.

Keep food and drinks away from the area your guests will be sitting. It is too easy to forget and grab a cracker. Chewing and drinking on playback can sound shocking when ten times louder

Collect the questions you want your guests to ask.

Choose a location. EVPs can record virtually everywhere; choose a quiet place.

Set up the Trigger Objects within the eye of cameras or video equipment. You will learn more about Trigger Objects later.

Set up a digital recorder. Many digital recorders have a selection for quality. Always choose the high-quality (HQ) or extra high quality (XHQ) setting. (See your recorder's manual.) Make sure you put in fresh batteries. Have EXTRA BATTERIES near. Every time we have documented the paranormal, all the batteries drain even when we have not collected EVPs, keep extras and place them away from recording devices, just in case.

The smell in the room may also act as a trigger, and scents can and have appeared during documenting. We ask guests not to wear perfume or highly scented lotions. We use a candle warmer and have experimented with different home smells; some of the favorites are cookies, pine trees, and apple cinnamon.

Have all guests bring a sweater or cloth jacket. In our experiences, the temperature can and does change drastically during communication. If coats are not cloth, the sounds they make can ruin the EVPs

Do not be concerned that you won't remember this. There is a Step by Step list at the end of this book: Timers and virtual clocks and timer apps for iPads and iPhones.

Always Play Nice

RULE #5
USE OUR TOOLS

In this chapter, equipment pictures are on the website.
Open QR Code when Shown to See Equipment In Action
You may have seen television shows that use specific equipment to communicate with spirits. Many of those devices work incredibly well; others do not.

Here we will tell you what works
Demonstrate how it works
Inform you of inexpensive alternatives to high priced items
Finally, tell you what we would never buy again!

Before running to your computer to buy anything, keep this in mind. To start checking your own house, you only need a few things.

EQUIPMENT MUST-HAVE ITEMS:
Digital voice recorder -two if possible
Camera or cell phones for stills/videos
Headphones or earphones-The best you can afford
Questions on paper (or from our website) ready in advance
Blank paper and pens for notes
30-second timer –(Free Apps available QR Code in Step by Step Chapter)

ITEMS WE USE & RECOMMEND:
2 Digital Voice recorders
Camera for stills/video-You can use your cell phone
Headphones or earphones-The best you can afford
Questions on paper (or from our website) in advance
Blank paper and pens
30-second timer (we use a sand timer)
The Rem-Pod
The Rook
THE SLS Camera
EMF Meters
Cat Balls
Boo Buddy
The Ovilus
The Paranormal Puck

EQUIPMENT DETAILS: I have included everything we tried that provided good results. All the equipment used is optional, and we do not use everything for every session. Hopefully, this section will help you if you want to purchase any of these devices.

EQUIPMENT DETAILS: To see photos and descriptions. See RULE 5 on the Website

EMR METER: EMR-ELECTROMAGNETIC & RADIATION Detects Electromagnetic energy. Human interaction cannot set it off. See the video of this device detecting EMF from a bone we found in the backyard

WATCH VIDEO: THE BONE

THE ORIGINAL REM-POD: Radiating Electromagnetism Pod is different from other tools, emitting its electromagnetic field. The alarm and lights will be activated when someone breaks the field.

WATCH: BOY TEN YEARS OLD- Part ONE & TWO
SCAN THIS NEW QR CODE:

THE REM-POD-2: This new REM-POD has a valuable function called ATDD, or Ambient Temperature Deviation Detection. The pod will make either a high or low pitched sound to indicate a sudden change in the air temperature and a visual signal - red or blue light, and it does even more. I heard something that came from the Rem-Pod itself.

WATCH THIS VIDEO PAGE: THE DAY THE REM-POD WENT CRAZY

SCAN THIS NEW QR CODE:

THE ROOK: The Rook is an EMF Meter with sound, Detecting small changes in electromagnetic energy and alerts with lights and sound.

MEL METER-EMF: Simultaneous EMF and Temperature Readout with Red Illuminated Display. Human interaction can not set it off.

K-2 METER: EMF and Temperature Readout with Red Illuminated Display. Human interaction can not set it off.

THE ROOK, THE MEL-METER, & K-2 METER, All Devices are used in the following videos:

WATCH VIDEOS: SCARED OF GOD- PART ONE & TWO
SCAN THIS NEW QR CODE- KEEP OPEN:

MEL METER-EMF 2: EMF, REM, and ATDD; Simultaneous EMF, EM Field output, Temp Alert with Illuminated Display and flashlight. Human interaction can not set it off.

BOO BUDDY: Detects: Electromagnetic energy, temperature changes, sounds, has sensors for touch, and talks to attract EVPs - He is not just a bear- he is your partner in investigating. Boo Buddy is the best for the money, Equipment, and Trigger objects.

WATCH: BOO-BUDDY
Same QR CODE

THE SPIRIT BOX - SB7T: A "Spirit Box" (also commonly referred to as Ghost Boxes) is a machine that picks up the verbal communications of spirits. From handheld AM/FM radios that will not stop when you hit a particular frequency. As a result, the radio will continually jump from station to station so fast that you know it must be paranormal if you hear more than one syllable phrase.

THE OVILUS: The Ovilus, or Puck, is an electronic speech-synthesis device that utters words depending on electromagnetic waves in the air using an EMF Meter. The device is by Bill Chappell, a retired electronics engineer interested in the paranormal who makes devices for the field.

This video will demonstrate both: THE SPIRIT BOX & THE OVILUS,

WATCH: THE ESTES METHOD
Same QR Code

THE PARANORMAL PUCK: The Paranormal Puck 2 will monitor EMF, temperature, humidity, light levels, barometric pressure, movement, and ionization. The app will allow all data to be seen, visualized, and shared directly from your smartphone or tablet, compatible with iOS and Android.

WATCH: THE ECHO-VOX
Same QR Code

The Echo-Vox is an app that works best on an iPhone or iPad. Links to download are available in Chapter STEP BY STEP

CAT BALLS: Yeah, that's right! A Simple Cat Toy works excellent. The slightest movement starts the lights. It's easy to use and does not make a sound. Puuuurrrrfect tool for investigating

A PARABOLIC MICROPHONE: A microphone uses a parabolic reflector to collect and focus sound waves onto a transducer, in much the same way that a parabolic antenna (e.g., satellite dish) does with radio waves. Though they lack high fidelity, parabolic microphones have great sensitivity to sounds in one direction, along the dish's axis, and can pick up distant sounds. Typical uses of this microphone include nature sound recordings such as recording bird calls, field audio broadcasting and eavesdropping on conversations, such as espionage and law enforcement.

WATCH: BOO & ROOK
Same QR Code

SLS CAMERA: The handheld SLS camera was my favorite tool, but that changed after our daughter and son-in-law made a better version. Instead of a tablet, they used a whole computer. The camera is attached to the computer with a very long connection, and the cord is so long I can still walk around. The added benefit is that batteries die very fast; that is resolved and works great! Many videos demonstrate how well it captures phenomena. If you are interested in having one made for you, it may be possible contact us. It uses an RGB camera with depth sensor and an infrared light projector with a monochrome CMOS sensor which sees everything not as a flat image but as dots arranged in a 3D environment. 1000's infrared dots allow the camera to "see" depth and detail like a sonar.

WATCH: SLS CAMERA
Same QR Code

CAMERAS: Almost any camera can photograph something paranormal. See the many types of cameras that can you can use.

PHOTOGRAPHIC EVIDENCE in GUEST MODE- This does not require a QR Code:
www.theirrules.com/photographicevidencepage1

OPEN: PHOTOGRAPHIC EVIDENCE Members Only
SCAN THIS NEW QR CODE – KEEP OPEN:

There is a secret to photographing and recording orbs.
WATCH ORBS; YOU CAN DO THIS
Same QR Code as above

+

DON'T BUY THIS: We have purchased many items, but they did not all do as promised. Some function correctly but don't help in an investigation. A typically used device is a motion sensor alarm. It works great, but the sound is so loud it disrupts any EVP.

These items may work in other homes; these devices contaminate any research in our home.

SEE WHAT DEVICES NEVER WORKED:
SCAN THIS NEW QR CODE:

WATCH THE SOUND: EVP's: Electronic Voice Phenomena is more commonly known as EVP

Set up the recorder. Many digital recording devices have a selection for quality. Always choose the high-quality (HQ) or extra high quality (XHQ) setting. Make sure you put in fresh batteries and have extra batteries close.

EVP Recordings (in GuestMode) vary by gender (men and women), age (adults and children), tone, emotion, and language, and some are more easily heard and understood than others.

No QR Code required: www.theirrules.com/e-v-p-s

Most EVPs consist of single words, phrases, or short sentences, although sometimes, they are of grunts, groans, growling, or other vocal expressions. Sometimes the voices can be heard immediately; other times, they are only heard in playback.

WATCH: GHOST BOX PART 1
SCAN THIS NEW QR CODE KEEP OPEN:

Sometimes the EVP will require editing software for amplification or a slowing down to listen to it. The quality of EVP also varies. Some are difficult to distinguish, with meanings that are open to interpretation. Some EVPs, however, are easy to understand. EVP often has an electronic or mechanical character, although sometimes, it can be very natural sounding.

THE QUALITY OF EVPs is as follows:

Class A:
Easily understood. These are also usually the loudest EVPs

Class B:
Warping of the voice in certain syllables. Lower in volume

Class C:
Excessive warping. They are the lowest in

During one of our playbacks, one of our guests became very frightened due to the language.
The EVP response was, "I don't fuckin care" another was, "Who the fuck are you?"
When she first heard the man's gruff voice, she interpreted it as hostile. After thinking about it, we all realized so many people use the F word in almost every sentence they say. Perhaps this man was just one of those people. He never said the F word directly to us; he never said, "Hey, fuck you."
As far as the yelling in EVP, we did ask them to talk as loud as possible, so keep that in mind. Lastly, we turn up the sound after putting the EVP through editing software.

Sometimes a quiet grunt or a Class B EVP with the volume up can sound terrifying, and other times they can be easy to miss. Listen carefully, or you may delete an EVP you want to hear.

LISTEN TO: THE GIGGLE
Same QR Code

TAGGING: It is helpful to record a quiet session the night before and log the sounds you hear. We have trained our brains to filter out a lot of background noise in our everyday life, but your recorder will pick up everything. You will be surprised what normal sounds sound like a woman screaming on playback. Once you know those noises, use tagging and name them during your recording to not be mistaken for EVP.

Anytime a group of people get together, someone will cough, sneeze or clear their throat. While you are recording, appoint someone as the "Tagger" to mark the sound in the recording. "Dog barking," "A car." "Judy coughed"

Always try to keep the "TAG" as short as possible and speak in normal tones. Never Whisper A whisper on playback will confuse the listener since it almost always sounds like an EVP. Because we used Tagging in this recording on playback, we could hear a phantom dog. If dogs had been in the house, we would have tagged

"Dog."

WATCH: UNEXPLAINED DOG
Same QR Code

PLAYBACK ON-DEMAND: Choose someone in your group to handle the playback recorder. You are trying to pick up voices that can often be soft, subtle, and hard to hear, so keep the environment as quiet as possible. You have already shut off the radios, TVs, computers, and any other sources of extraneous noise. Now listen and record the room. Something happened to me; I discovered we have two lightbulbs in a seven-bulb chandelier that make a sound. I would never have noticed this if I did not record a soundcheck before. Doing a good recording test prior is a great way to check your refrigerator, coffee maker, and other appliances in your home that make noises. Avoid moving around to eliminate the sounds of footsteps and the rustling of clothing. Audio Pareidolia can and does happen.

LEAVE THE ROOM: Another recording option is to place a recorder in a room and leave. State your name, place, and time, then set the recorder down and leave the area. After fifteen or

twenty minutes, retrieve your device and listen to what you may have captured. The disadvantage of this method is that you aren't present to hear and discount any ambient noises. Even if you stay in the room with your recorder, it's best to set the recorder and microphone down on something like a chair or table to eliminate the possible noise of your hands on the devices.
(This EVP) I turned on by my front door and left. It sounds like someone is touching it.

LISTEN TO: WHEN NO ONE IS HOME
Same QR Code

AUDIO PAREIDOLIA: When you hear something in the playback recorder, remain open-minded to the possibility that what you are hearing could be The Pareidolia effect. Typically described as a visual phenomenon, it can also be auditory. There have been times when I wanted to hear an answer so badly; I hear it when no one else could. Have another person listen to the recording without telling them what you hear. My standard of documenting is that if three people hear what I heard, I mark it as such. Have others listen to confirm what you are hearing

EVP CAN WORK BOTH WAYS: I have a recording of EVPs I collected in one night. I have noticed that it increases the activity when I play it before documenting the house. Please be warned that this EVP recording has disturbed some people; if you become overwhelmed, turn it off.

WATCH: EXPLANATION OF TRIGGER EVP FIRST
Same QR Code

WARNING: WARNING: WARNING: WARNING:

Please be warned that the TRIGGER EVP recording has disturbed some people; Please watch the previous warning video before watching this TRIGGER EVP.

WARNING: WARNING: WARNING: WARNING:
If you become overwhelmed, turn it off.

LISTEN TO: TRIGGER EVP
Same QR Code

EDITING YOUR EVPs: There will be times when your recording has so much White Noise that the EVP is almost impossible to understand. A program like " Audacity" will allow you to capture a section of the white noise; this is accomplished by listening to an area with no voices but is just static. Follow the directions in the program to copy and then remove just that sound from the entire recording.

Once you hear the voices, you can increase the volume, and it may take practice since the volume increase will bring back the white noise.
This software lets you boost low volume, eliminate some background noise, and add echos to listen repeatedly. It will allow you to cut out the specific EVP sections of the recording and delete the sections of the recording that have nothing of value.

Again, this just takes practice.
There are other programs available to download on your computer.

However, "Audacity" at this time is free. Access to Links in STEP BY STEP

TRUST YOURSELF: If at any time you feel uncomfortable, stop. No one has any expertise in this area.
The best thing to count on is your gut feelings.

RULE #6
WE WANT TO PLAY WITH THAT

Trigger Objects are one of the most effective methods to draw attention to an area or room.
(Guest Mode www.theirrules.com/rule6)

Choosing the trigger objects that give you the best results will take time. If you have no information about age or gender, you can add various attractive things to test the results—jewelry, toys, music box, etc.

OLDER TOYS Sometimes, objects can draw attention without any real interaction with the item. Children's toys can sometimes work with any age. Think about the toys you played with as a child. We have attracted various EVPs and photographic evidence when one of these toys is in the room.

Perhaps they recognize the item and wish to take a closer look. Even something like an old phone can trigger a response.

VIEW THE SLIDESHOW for inexpensive alternatives to very pricey paranormal Triggers as seen on tv

SCAN THIS NEW QR CODE KEEP OPEN:

A recent example was OLDER TOYS; Debbie; I set out an Etch-a-Sketch and a Simon. Suddenly an adult-looking figure appeared on the SLS Camera and later, a child. (The Ovilus had announced her name was Debbie.)

WATCH: OLDER TOYS DEBBIE
Same QR Code

UNEXPECTED TRIGGERS: Remember that we are not setting these things out with expectations that they will move around the room right before us. We have had things move in the house, but the goal is simply attraction in this case. We have had many unexpected things act as Trigger Objects:

WATCH: THE CROW
WATCH: BOO-BUDDY
WATCH: THE BONE
Same QR Code

INEXPENSIVE ALTERNATIVES: Many paranormal websites will sell Trigger Objects (some at a very high price). We have discovered an unlimited variety of everyday items that can work. Each person is different, and what works for us might not work for you. We have also found ways around the high-priced equipment while still achieving the same effect.

OPEN RULE 6 (GuestMode): www.theirrules.com/rule6
Our recordable (You can also record the music of your choice) door chime at the cost of 4.00 online can replace an item sold for sold that has sold for over 300.00. The door chime does not have the spooky look of the expensive one, but it's also typically used in the dark.

The most used Trigger Object in our home is a small crow our daughter had bought me as a stocking stuffer for Christmas. One afternoon I heard a sound coming from the living room and investigated. The crow was playing on its own. The small, manageable action button that plays "NeverMore" This crow, never intended to be a Trigger Object, happily fascinates a few of the children we have recorded inside the house.

It will be fun to see what items will act as Trigger Objects inside your home. There is no guarantee of initial attraction, and at times objects ignored one day will be popular the next.

FREE: There are also many free options for Trigger/Equipment alternatives.

The WHITEBOARD: I installed mine on an IPAD, and leaving it on provides the opportunity for them to mark the board with even the slightest touch. I added a child's coloring book page and set the board; I asked them to color the kids

Trigger Objects can be almost anything. See the list of everyday items and inexpensive replacements for high-cost equipment.

Some investigators have found Trigger Objects to be the cause of a haunting.

Commonly Used Trigger Objects:

Books/Bibles
Cigars / Cigarettes
Childhood Toys
Jewelry
Music Box –(also in replacements)

Trigger Objects We Discovered by chance:

Birthday Cake /Wrapped Gifts
Christmas Tree / Ornaments
Decorations /Home /Holiday
Modern 3D light
Door Chime*
Nostalgic Items and Toys
Religious Symbols
Scented Candle Wax
Televisions – Computers – iPads – Cell Phones

The smell in the room may also act as a trigger; we use a candle warmer and have experimented with different home smells; some of the favorites are cookies, pine trees, and apple cinnamon.

OTHER TOYS: Some equipment used for Paranormal research can also act as a trigger object. Recently Laurel and I were sitting and talking, and I almost always leave on some piece of equipment. This particular day we left on the Rem-Pod. I asked Laurel if she wanted to play a game, and the Rem-Pod went off. That's how it happens sometimes. While leaving on digital recorders, we have also heard them comment on our conversations. As if they are always listening and suddenly hear something that interests them.

The game we played was The Ouija Board app downloaded on my iPad. While playing the game, the Rem-Pod signaled answers.

GUEST MODE does not require a QR Code.
See the Videos. It is AMAZING!
www.theirrules.com/triggerobjectsinactionpage1

RULE #7
ARE WE WHAT YOU EXPECTED?

Have you watched movies and television shows where the evil spirit demands that the homeowner leave the house?
It is best to try NOT TO HAVE EXPECTATIONS.

There has been such a stigma on the paranormal that it will take a while for these preconceived ideas to change. It is difficult to change how we think about something, and our beliefs are very personal. The media constantly portraying the paranormal as evil will sell movie tickets, but that doesn't make it accurate.
Still a part of being human, you will feel scared sometimes.
Since I moved into this home, there is not one thing I can say resembles any movie I have ever seen. Many of the events in this house I had never heard of before. For instance, our light bulbs unscrew themselves; batteries will go dead and then come back to full strength.

In the last ten years, I have learned that those who reside in our home with us are frightened, confused, and crying out for help. Empathy and respect toward the dead should be the goal.
We have seen many unexplainable and strange things here, but we have learned even more. We want to help others understand the reality of what to expect.

You may have heard of the term INTENTION MATTERS; I believe that. I have been afraid in my house, but I do my best to stop and think about it logically. If they wanted to hurt me, they would have an easy time of it. I am partially disabled and have a difficult time walking downstairs. (I feel blessed I can still do it) if they wanted to hurt me, they could blow on me and push me down the stairs.

My intention is not to hurt them, and their intention is not to harm me.

The fact that this is not scary does not take any of the thrills out of it. I sit in awe, watching what is unimaginable. When we hear the water go on by itself, the lights start to flash or listen to sounds we can't explain; I turn on a recorder and sometimes capture EVP's asking for help. Other times they seem to be just lonely and want someone to talk to; I'm not sure if they can all see and talk with each other. I have captured two unseen people together on the SLS camera and recorded many Unseen People talking to each other;

WATCH: THE COUPLE
SCAN THIS NEW QR CODE KEEP OPEN:

Other times it seems like some of them are all alone and very scared.

I have had people tell me they think this is crazy, so what's more crazy? Believing demons are in your house or that it may be a benign spirit who wants someone to talk with them? I named this chapter "Are We What You Expected" because if I were going to write to you on behalf of them, I believe they would say this;

I deserve respect and dignity. Please don't hurt my feelings
I am not a joke; my life was as hard or harder than yours.
I exist whether you want to acknowledge me or not.
I want to help, not hurt. I won't understand everything.
I am not a circus act; I will not respond for your entertainment.
I get mad just like you; that does not make me evil.

Please be kind. I might be a child, a grandparent, a friend, or even one of your moved-on relatives.

WE CAN BE ANYWHERE

When we first started documenting the activity, we focused on debunking the events, not researching. We had purchased some equipment we had seen on the popular ghost shows.

The SLS camera intrigued me. We bought a handheld version with a tablet for the screen (now we have a computer version). The first day I used it, I planned to walk around the house and see if we could document anything. I opened my bedroom door, and a significant seven to eight-foot green stick figure was right in front of me as if it had been listening at the door. I decided it was a glitch in the program and moved on.

My husband and I took our new SLS Camera outside to see if the stick figures were fake. At this time, I fluctuated from believer to skeptic almost daily. We walked around with the camera and saw nothing at first. In the garden, we have a pole, and a stick figure seemed to hide there. Of course, we determined the camera was making something out of nothing. We kept walking and soon saw another figure who appeared to be floating by and almost directed us; we walked in that direction once the figure disappeared.

This video is long, but I feel it's worth watching to the end, where the person we capture on video holds out his hand to stroke my husband's face; I almost cried.

An Unforgettable Day: Attempting to assist a spirit in trouble
WATCH: OUTSIDE SPIRIT
Same QR Code

I used the SLS camera at least once a week, and I found it fascinating. There were still times I was sure the camera might be creating figures out of household items that were similar in shape to a person. In the house, our staircase has a ball on the top of the railing, and Jim and I both thought it was creating the stick figures out of their shape. (Banister: the structure formed by the uprights and handrail at the side of a staircase.)

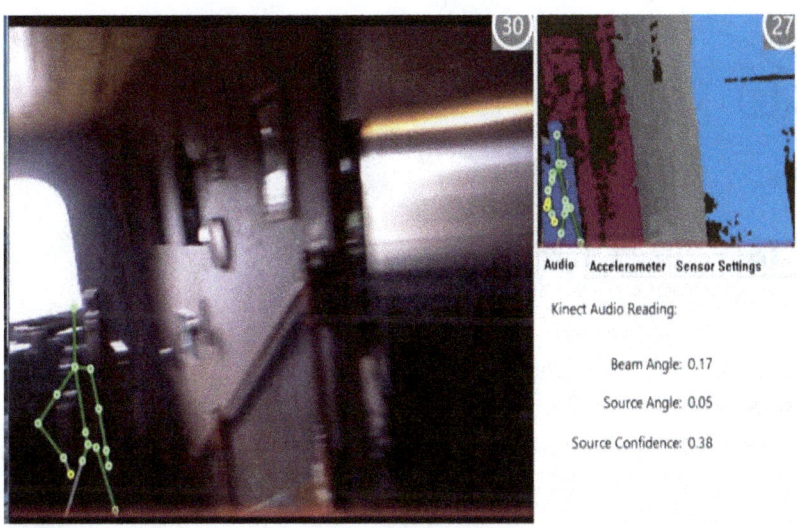

While watching the staircase one day, a figure appeared, and then jumped away from the staircase and started to walk into the living room. That answered that!
Many times we have recorded them just popping in. In this video, one of the children grabs onto me.

WATCH: CHILD GRABS ONTO ME
Same QR Code

So why is this a rule? Imagine what they hear if you're talking about them if they could see us and listen to us. When I first started documenting the activity inside the house, I had many misconceptions about what to ask, say, and what was happening. One of my mistakes was that before and after recordings, I would talk as if the unseen were not there. In my mind, I guess I assumed once the recorder went off, they could not hear us anymore. Well, that's not true.

They can be anywhere; Now, I firmly believe they are everywhere. Not just in a house or abandoned building. As you saw, they are just outside the front door.

Why are there so many inside our home? We have no idea. There are no records we have been able to find, so, unlike the television shows, we have no information that makes us all say, "Oh, now we understand." There is no evidence that this is not in every home. One thing that makes our house unique is that we investigate and document weekly, if not more.

This video will make you think twice:
WATCH- THE BATHROOM
Same QR Code

Even when they do not want to talk to you, they seem to be there. As the Rem-Pod indicates by the temperature meter going off, there is someone there. After I asked to speak to them, they backed away.

WATCH: REM-POD 2
SCAN THIS NEW QR CODE:

Something is attracting them to this house. It may be something to do with post-traumatic stress or PTSD, which I have; I have spoken to a few people from the PTSD and CPTSD Facebook groups who also experience paranormal activity during a PTSD episode. It may be the heightened emotion involved or the disorder itself; we may never know. I wrote a book about this subject due to my experiences here. When a PTSD event happens, the paranormal happens as well.

These are videos taken at Laurel's house; she was in the process of redecorating a spare room into her podcast room.

WATCH: A RESIDENT AT LAUREL'S HOUSE
WATCH: IS IT PETER OR DEREK?
WATCH: EVERYTHING IS GOING OFF
Same QR Code

Everyone does seem to agree that redecorating or remodeling can create activity, but no one knows why. (We can only guess)

CHECK EVERYTHING: After you start documenting the paranormal, check everything! I have had Videos and EVPs I thought held nothing but found it later.

Photos I had taken contained something unexplained, and I never noticed it until later.
SCAN THIS NEW QR CODE:

WATCH FOR THIS: If you are a person who watches those paranormal shows (I'm not allowed to say names), Watch for this: When they show an event about to happen, look for an orb right before they demonstrate the paranormal activity. I used to watch some of those shows and thought, well, that's ridiculous. As I became more open-minded, I noticed that you would see an orb in some of these videos right before something happened. I am starting to believe that an orb that flies by before a door opens; could be the energy of the entity opening the door.
Before a light and sound device goes off Rem-Pod, etc., an orb will pass first at our house. Depending on the camera type and setting. (most but not all cameras can photograph orbs - from RULE #5)

THE ORBING HOUR: At night, the house usually seems to come alive at 1:30-2:30 am. If these orbs are dust, and of course, some could be, but then what the hell is stirring them up? Why is it particular times? No one knows what Orbs are, how much of it is dust, and if anyone does know, please tell me how they can come through a window.

WATCH: ORBS IN ACTION PAGE
SCAN THIS NEW QR CODE:

It can already be spooky when you investigate your own house because you have to live there when it's over. It may be hard to accept this, but everything you discover when you conduct your first investigation was already there.

CHECK EVERYTHING:
An example is while editing the THEIR RULES website; I re-watched the *DEBBIE video and heard EVPs I never knew were there.

I NEVER LISTENED TO IT because I had placed my hand over the mic (accident).

THE FOLLOW-UP PAGE:
See the many things that are super easy to miss.

WATCH: THE FOLLOW-UP PAGE-
SCAN THIS NEW QR CODE:

ENTITY TOUCHES ROOK
THE GHOST BOX -PART TWO
DEBBIE UPDATED – EVPs I didn't hear before*
DRYER
ROOK & REM-POD

CHRISTMAS INTERVIEW – I was sent a message saying there were EVPs in the video, I still can't hear them. I suspect there may be EVPs in the clips from The Reliving.

OUTDOOR GAME CAMERA
SECURITY CAMERA- CAN YOU FIND IT?
ECHOVOX- ANOTHER DAY- Coming Soon

ABOUT THE AUTHORS

JUNE MATTHEWS

Born into an abusive and violent home, June was moved from parent to parent until age 12.
A runaway for years, she experienced horrific violence and abuse, followed by a lifetime of dysfunction and sadness until she entered an Alcohol and Drug treatment center. The treatment center changed her life.

She remained clean and sober while still struggling with memories from her past. Life and career changes led June and her husband to pack up their life in California and move into their dream house in an idyllic small town in Eastern Washington. While he traveled for work alone in their new home, peculiar events unfolded, and her life changed forever. A lifelong skeptic, June would never have imagined she'd become a paranormal researcher and author. What happened in the house forever changed her views on spirituality, the paranormal, and what it means to heal. She wanted to share her unique story with the desire that it would help others on their journey toward recovery and renewal.

Today she remains a private person, markedly restricted physically from injuries but no longer chained to her past. June and her husband Jim still live in the house, where she is hard at work on her next book, a sequel to THE RELIVING.
She continues her paranormal research and works on another book to share her techniques on Electronic Voice Phenomena. She loves entertaining guests in her home, taking pictures, creating jewelry with her daughter, crafting, decorating her house, and target shooting with her husband.
She is a proponent of The 12 Steps of Alcoholics Anonymous.

I visit her at the house frequently as it is hard for her to get out due to physical limitations; I often help her with whatever current home improvements she is working on (installing wood floors, tile) I am also present for one of June's favorite things, the yearly Halloween Party.

Written by,
Carolyn Martell

Find out more about June Matthews
www.AuthorJuneMatthews.com
www.TheirRules.com
www.TheReliving.com

LAUREL COOPER

My name is Laurel Cooper, and I am excited about this book.
I am 37 years old and have been dealing with the paranormal for as long as I can remember.
I was born in Beverly Hills and raised in California, surviving the earthquakes and riots. I graduated high school in 2001 and had a rough start to my adult life. I grew up and became a better person by learning from my mistakes. I moved to Spokane, Washington, ten years ago, met and married my wonderful husband and gained a son.

When I was younger, my experiences with the paranormal confused and frightened me.
I am currently writing a book about my supernatural struggles and what I learned.
After so many people telling me I had demons and was no good, one of my former bible teachers, an amazing man, said, "It's not demons; you are one of the gifted ones; I'm sorry that you have gone through this alone, but you are not a freak." I always felt like no one would believe me, but this man did, and he saved my life. For this reason, I have titled my upcoming book THE GIFTED ONES.
If you are interested in Paranormal Research outside your home, you may want to stick to abandoned buildings or inform the

homeowner that they can not be present. There is never a way to know what you will be walking into when you are in someone else's home, but you also never know how someone will react. If the homeowner has exposure and education with the paranormal, they will not need you there.

In my own experience, I have investigated a friend's house, I didn't believe I would find anyone in her home, but when I arrived there and started my investigation, it was apparent that there was someone else in the house with us. She instantly assumed it was her mother-in-law, and I had to remind her that "we can't know who this is."

Before the investigation, I told her that it was essential to try and be as quiet as possible. I warned her that if she felt uneasy, we should stop, and she was confident that wouldn't happen. As soon as I started, she was already making enough noise that possible EVPs were drowned out by her screaming or gasping.

I felt frustrated that this had wasted evidence and time for both me and the others. I did get some good evidence and was able to have an open dialogue with someone. Still, I wish I could have heard the entire session but now I will never know.
We don't know how hard it is for them (the others) to communicate, and we need to show them that respect.
I felt as if I should not have come and fought the desire just to leave her house. I finished what I could, packed up the equipment, and was relieved to go.
Once I was back home, she called me to tell me it was getting worse and she was afraid of being possessed. I went back to the house to help her calm down; she was scared and believed that someone (maybe a demon) was attacking her. I did not understand what she meant by attacking; I did not see anything wrong.
Minutes later, she was fine, and I never really knew if her intention

was genuine or not. I (cleansed the house) prayed for help, and made a final retreat.

I don't believe that her home had a demon. Regardless, she feels better, and the family said the house felt better.

It does not matter if it is the paranormal or someone you know; if things do not feel right, always trust your instinct; if you think it's time to leave, do that.

THE 7th SENSE PARANORMAL VODCAST WEBSITE
www.The7thSensePodcast.com

Contact The Seventh Sense Host, Laurel Cooper, for:

Authors
FilmMakers
Gamers
Hotel &Business Owners
Inventors
Influencers.

SEND ME YOUR PARANORMAL CLIPS-FOOTAGE-OR PHOTOS TO BE ON THE SHOW.

Promote your Book, Movie, Haunted Location, Invention, or share YOUR Paranormal Experience on

THE 7th SENSE PARANORMAL VODCAST.

We would love to have you on our show!

Find out more about

Laurel Cooper

www.The7thSensePodcast.com

www.TheirRules.com

STEP by STEP

Here is a step by step list to help you:

1st -Become a Member of The Website.

2nd -Read the Instructions In THEIR RULES

3rd -Read the THEIR RULES, and as you read the book, open the links to learn more.

4th -Choose the day and time of the event.

5th - Invite friends to help you

6th -Collect the items you need for a successful recording:

2 Digital Recorders - Use digital-only, not cell phones

Cameras - Cell phones will work

Headphones- noise canceling works best

Printed Questions (available free in this book)

Flashlight or Candles

30 Second Timer

7th -Run a recorded soundcheck through the room you will be recording.

8th -Before the event, ensure that your guests bring a cloth jacket/sweater –and Trigger Object for Plate.

9th - Before the event, decide on foods and drinks.

10th -On the same day of your event, run through the list:

PREPARE THE SPACE:

SERVE food an hour or more before recording.

TURN off things that create sounds, clocks, air conditioning, fans, house phone and have guests shut off ringers on their cell phones.

ITEMS with words like "Ghost Hunting" should be removed.

SET UP the room so that ALL your guests are comfortable. While recording a guest moving in their chair can sound incredibly confusing on playback.

COLLECT the questions you want your guests to ask.

SET UP the Trigger Objects within the eye of cameras or video equipment.

SET UP a digital recorder. Many digital recorders have a selection for quality. Always choose the high-quality (HQ) or extra high quality (XHQ) setting. (See your recorder's manual.) Make sure you put in fresh batteries. Have extra batteries near. Every time we have documented the paranormal, all the batteries drain even when we have not collected EVPs. Make sure to keep extras and place them away from recording devices, just in case.

SET UP timer and light, find virtual clocks and timer apps for iPads and iPhones.

11th -Check The WARNING LIST

IF YOU or GUEST:

Received terrible news,

Argued with someone on the same day

Feeling sick

Feeling depressed

Not had enough sleep

Been drinking

Using drugs

12th -Explain to your guests that they must be polite; this is a serious event but also meant to be fun. Perhaps have them all read RULE 2 the night they arrive.

13th -Make sure you have everything set up, recorder on, etc.

OPEN LINKS:

SCAN THIS NEW QR CODE:

For free 30 Second Timers

Inexpensive and Free Paranormal Programs

Free Editing Programs

Games as Trigger Objects and more.

NOTES

Date:_____

Time:_____

The Paranormal is REAL

Look for the hashtag:
#theparanormalisreal

The Reliving 1st Edition – Book & DVD

The Reliving 2nd Edition – Interactive Book QR

Their Meeting Place – Interactive Book QR – Coming Oct 2022 Sequel to THE RELIVING

The Dream Journals – Book – There is more than paranormal activity in our haunted house; there is also THE DREAMS!

SOCIAL MEDIA:

TIKTOK: The Paranormal is Real

THEIR RULES – Facebook – YouTube-

THE RELIVING – Facebook Fan Page:
www.facebook.com/thereliving

Facebook Private Group: Friends Who Like THE RELIVING

Facebook: Inside The Reliving House
facebook.com/insidetherelivinghouse

Facebook: Author June Matthews
www.facebook.com/authorjunematthews

Facebook: 7th Sense Laurel Cooper
www.facebook.com/the7thsenseparanormalpodcast

FIND US ON
Twitter – Instagram – TIK TOK
Snap Chat – LinkedIn -YouTube Pinterest

www.ingramcontent.com/pod-product-compliance
Lightning Source LLC
Chambersburg PA
CBHW050650160426
43194CB00010B/1888